Charles Kingsley

The Limits of Exact Science as Applied to History

Charles Kingsley

The Limits of Exact Science as Applied to History

ISBN/EAN: 9783337034085

Printed in Europe, USA, Canada, Australia, Japan

Cover: Foto ©berggeist007 / pixelio.de

More available books at **www.hansebooks.com**

BY THE SAME AUTHOR.

MISCELLANIES. Second Edition. 2 Vols. 18*s.*

TWO YEARS AGO. Third Edition. 6*s.*

WESTWARD HO! Third Edition. 6*s.*

THE SAINT'S TRAGEDY. Third Edition. 5*s.*

YEAST. Fourth Edition. 5*s.*

ALTON LOCKE. Fourth Edition. 2*s.*

HYPATIA. Third Edition. 6*s.*

PHAETHON. Third Edition. 2*s.*

ALEXANDRIA AND HER SCHOOLS. 5*s.*

GLAUCUS; OR THE WONDERS OF THE SHORE. Coloured Illustrations. Fourth Edition. 7*s.* 6*d.*

THE HEROES. Second Edition. Illustrated. 5*s.*

ANDROMEDA AND OTHER POEMS. Second Edition. 5*s.*

THE GOOD NEWS OF GOD. SERMONS. Third Edition. 6*s.*

SERMONS FOR THE TIMES. Second Edition. 3*s.* 6*d.*

VILLAGE SERMONS. Fifth Edition. 2*s.* 6*d.*

SERMONS ON NATIONAL SUBJECTS. First Series. Second Edition. 5*s.*

SERMONS ON NATIONAL SUBJECTS. Second Series. Second Edition. 5*s.*

THE LIMITS OF EXACT SCIENCE
AS APPLIED TO HISTORY.

Cambridge:

PRINTED BY C. J. CLAY, M.A.
AT THE UNIVERSITY PRESS.

THE LIMITS OF EXACT SCIENCE
AS APPLIED TO HISTORY.

An Inaugural Lecture,

DELIVERED BEFORE

THE UNIVERSITY OF CAMBRIDGE.

BY THE REV.

CHARLES KINGSLEY, M.A.

PROFESSOR OF MODERN HISTORY IN THE UNIVERSITY OF CAMBRIDGE,
CHAPLAIN IN ORDINARY TO THE QUEEN, & RECTOR OF EVERSLEY.

MACMILLAN AND CO.
Cambridge:
AND 23, HENRIETTA STREET, COVENT GARDEN,
London.
1860.

TO

THE UNDERGRADUATES

OF THE

University of Cambridge.

THE LIMITS OF EXACT SCIENCE AS APPLIED TO HISTORY.

It is with a feeling of awe, I had almost said of fear, that I find myself in this place, upon this errand. The responsibility of a teacher of History in Cambridge is in itself very heavy: but doubly heavy in the case of one who sees among his audience many men as fit, it may be some more fit, to fill this Chair: and again, more heavy still, when one succeeds a man whose learning, like his virtues, I can never hope to equal.

But a Professor, I trust, is like other men, capable of improvement; and the great law, *docendo disces,* may be fulfilled in him, as in other men. Meanwhile, I can only promise that the whole of such small powers as I

possess will be devoted to this Professorate;
and that it will be henceforth the main object
of my life to teach Modern History after a
method which shall give satisfaction to the
Rulers of this University.

And I shall do that best, I believe, by
keeping in mind the lessons which I, in common with thousands more, have learnt from
my wise and good predecessor. I do not
mean merely patience in research, and accu-
racy in fact. They are required of all men:
and they may be learnt from many men. But
what Sir James Stephen's life and writings
should especially teach us, is the beauty and
the value of charity; of that large-hearted
humanity, which sympathizes with all noble,
generous, earnest thought and endeavour, in
whatsoever shape they may have appeared;
a charity which, without weakly or lazily
confounding the eternal laws of right and
wrong, can make allowances for human frailty;
can separate the good from the evil in men

and in theories; can understand, and can forgive, because it loves. Who can read his works without feeling more kindly toward many a man, and many a form of thought, against which he has been more or less prejudiced; without a more genial view of human nature, a more hopeful view of human destiny, a more full belief in the great saying, that "Wisdom is justified of all her children"? Who, too, can read those works without seeing how charity enlightens the intellect, just as bigotry darkens it; how events, which to the theorist and the pedant are merely monstrous and unmeaning, may explain themselves easily enough to the man who will put himself in his fellow-creatures' place; who will give them credit for being men of like passions with himself; who will see with their eyes, feel with their hearts, and take for his motto, "Homo sum, nil humani a me alienum puto"?

I entreat gentlemen who may hereafter

attend my lectures to bear in mind this last saying. If they wish to understand History, they must first try to understand men and women. For History is the history of men and women, and of nothing else; and he who knows men and women thoroughly will best understand the past work of the world, and be best able to carry on its work now. The men who, in the long run, have governed the world, have been those who understood the human heart; and therefore it is to this day the statesman who keeps the reins in his hand, and not the mere student. He is a man of the world; he knows how to manage his fellow-men: and therefore he can get work done which the mere student (it may be) has taught him ought to be done; but which the mere student, much less the mere trader or economist, could not get done; simply because his fellow-men would probably not listen to him, and certainly outwit him. Of course, in proportion to the depth, width, soundness,

of his conception of human nature, will be
the greatness and wholesomeness of his power.
He may appeal to the meanest, or to the lof-
tiest motives. He may be a fox or an eagle;
a Borgia, or a Hildebrand; a Talleyrand, or a
Napoleon; a Mary Stuart, or an Elizabeth:
but however base, however noble, the power
which he exercises is the same in essence.
He makes History, because he understands
men. And you, if you would understand
History, must understand men.

If, therefore, any of you should ask me
how to study history, I should answer—
Take by all means biographies : wheresoever
possible, autobiographies; and study them.
Fill your minds with live human figures;
men of like passions with yourselves; see
how each lived and worked in the time and
place in which God put him. Believe me,
that when you have thus made a friend of
the dead, and brought him to life again, and
let him teach you to see with his eyes, and

feel with his heart, you will begin to understand more of his generation and his circumstances, than all the mere history-books of
the period would teach you. In proportion
as you understand the man, and only so, will
you begin to understand the elements in
which he worked. And not only to understand, but to remember. Names, dates, genea-
logies, geographical details, costumes, fashions,
manners, crabbed scraps of old law, which
you used, perhaps, to read up and forget again,
because they were not rooted, but stuck into
your brain, as pins are into a pincushion, to
fall out at the first shake—all these you will
remember ; because they will arrange and
organize themselves around the central human figure : just as, if you have studied
a portrait by some great artist, you cannot
think of the face in it, without recollecting
also the light and shadow, the tone of
colouring, the dress, the very details of the
background, and all the accessories which

the painter's art has grouped around; each with a purpose, and therefore each fixing itself duly in your mind. Who, for instance, has not found that he can learn more French history from French memoirs, than even from all the truly learned and admirable histories of France which have been written of late years? There are those, too, who will say of good old Plutarch's lives, (now-a-days, I think, too much neglected,) what some great man used to say of Shakspeare and English history—that all the ancient history which they really knew, they had got from Plutarch. And I am free to confess that I have learnt what little I know of the middle-ages, what they were like, how they came to be what they were, and how they issued in the Reformation, not so much from the study of the books about them (many and wise though they are), as from the thumbing over, for years, the semi-mythical saints' lives of Surius and the Bollandists.

Without doubt History obeys, and always has obeyed, in the long run, certain laws. But those laws assert themselves, and are to be discovered, not in things, but in persons; in the actions of human beings; and just in proportion as we understand human beings, shall we understand the laws which they have obeyed, or which have avenged themselves on their disobedience. This may seem a truism : if it be such, it is one which we cannot too often repeat to ourselves just now, when the rapid progress of science is tempting us to look at human beings rather as things than as persons, and at abstractions (under the name of laws) rather as persons than as things. Discovering, to our just delight, order and law all around us, in a thousand events which seemed to our fathers fortuitous and arbitrary, we are dazzled just now by the magnificent prospect opening before us, and fall, too often, into more than one serious mistake.

First; students try to explain too often all the facts which they meet by the very few laws which they know; and especially moral phænomena by physical, or at least economic laws. There is an excuse for this last error. Much which was thought, a few centuries since, to belong to the spiritual world, is now found to belong to the material; and the physician is consulted, where the exorcist used to be called in. But it is a somewhat hasty corollary therefrom, and one not likely to find favour in this University, that moral laws and spiritual agencies have nothing at all to do with the history of the human race. We shall not be inclined here, I trust, to explain (as some one tried to do lately) the Crusades by a hypothesis of overstocked labour-markets on the Continent.

Neither, again, shall we be inclined to class those same Crusades among "popular delusions," and mere outbursts of folly and madness. This is a very easy, and I am sorry

to say, a very common method of disposing
of facts which will not fit into the theory, too
common of late, that need and greed have
been always, and always ought to be, the
chief motives of mankind. Need and greed,
heaven knows, are powerful enough: but I
think that he who has something nobler in
himself than need and greed, will have eyes
to discern something nobler than them, in
the most fantastic superstitions, in the most
ferocious outbursts, of the most untutored
masses. Thank God, that those who preach
the opposite doctrine belie it so often by a
happy inconsistency ; that he who declares
self-interest to be the mainspring of the world,
can live a life of virtuous self-sacrifice; that
he who denies, with Spinoza, the existence of
free-will, can disprove his own theory, by wil-
ling, like Spinoza, amid all the temptations of
the world, to live a life worthy of a Roman
Stoic ; and that he who represents men as the
puppets of material circumstance, and who

therefore has no logical right either to praise
virtue, or to blame vice, can shew, by a healthy
admiration of the former, a healthy scorn of
the latter, how little his heart has been cor-
rupted by the *eidola specûs*, the phantoms of
the study, which have oppressed his brain.
But though men are often, thank heaven,
better than their doctrines, yet the goodness
of the man does not make his doctrine good;
and it is immoral as well as unphilosophical
to call a thing hard names simply because
it cannot be fitted into our theory of the
universe. Immoral, because all harsh and
hasty wholesale judgments are immoral; unphilosophical, because the only philosophical
method of looking at the strangest of phænomena is to believe that it too is the result of law, perhaps a healthy result; that
it is not to be condemned as a product of
disease before it is proven to be such; and
that if it be a product of disease, disease
has its laws, as much as health; and is a

subject, not for cursing, but for induction;
so that (to return to my example) if every
man who ever took part in the Crusades
were proved to have been simply mad, our
sole business would be to discover why he
went mad upon that special matter, and at
that special time. And to do that, we must
begin by recollecting that in every man who
went forth to the Crusades, or to any other
strange adventure of humanity, was a whole
human heart and brain, of like strength and
weakness, like hopes, like temptations, with
our own; and find out what may have driven
him mad, by considering what would have
driven us mad in his place.

May I be permitted to enlarge somewhat
on this topic? There is, as you are aware,
a demand just now for philosophies of History.
The general spread of Inductive Science has
awakened this appetite; the admirable con-
temporary French historians have quickened
it by feeding it; till, the more order and

sequence we find in the facts of the past, the more we wish to find. So it should be (or why was man created a rational being?) and so it is; and the requirements of the more educated are becoming so peremptory, that many thinking men would be ready to say (I should be sorry to endorse their opinion), that if History is not studied according to exact scientific method, it need not be studied at all.

A very able anonymous writer has lately expressed this general tendency of modern thought in language so clear and forcible that I must beg leave to quote it:—

"Step by step," he says, "the notion of evolution by law is transforming the whole field of our knowledge and opinion. It is not one order of conception which comes under its influence: but it is the whole sphere of our ideas, and with them the whole system of our action and conduct. Not the physical world alone is now the domain of inductive science, but the moral, the intellectual, and the spiritual are being added to its empire. Two co-ordinate ideas pervade the vision of every thinker, physicist or moralist, philosopher or priest. In the physical and the moral world, in the natural and the human, are ever seen two forces—invariable rule, and continual advance; law and action; order and progress; these two powers working harmoniously together, and the result, inevitable sequence, orderly movement, irresistible growth. In the physical world indeed, order is most prominent to our eyes; in the moral world it is progress, but both exist as truly in the one as in the other. In the scale of nature, as we rise from the inorganic to the organic, the idea of change becomes even more distinct; just as when we rise through the gradations of the moral world, the idea of order becomes more difficult to grasp. It was the last task of the astronomer to show eternal change even in the grand order of our Solar System. It is the

crown of philosophy to see immutable law
even in the complex action of human life. In
the latter, indeed, it is but the first germs
which are clear. No rational thinker hopes
to discover more than some few primary
actions of law, and some approximative theory
of growth. Much is dark and contradictory.
Numerous theories differing in method and
degree are offered; nor do we decide between
them. We insist now only upon this, that
the principle of development in the moral, as
in the physical, has been definitively admitted;
and something like a conception of one grand
analogy through the whole sphere of knowledge, has almost become a part of popular
opinion. Most men shrink from any broad
statement of the principle, though all in some
special instances adopt it. It surrounds every
idea of our life, and is diffused in every branch
of study. The press, the platform, the lecture-
room, and the pulpit ring with it in every
variety of form. Unconscious pedants are

proving it. It flashes on the statistician through his registers; it guides the hand of simple philanthropy; it is obeyed by the instinct of the statesman. There is not an act of our public life which does not acknowledge it. No man denies that there are certain, and even practical laws of political economy. They are nothing but laws of society. The conferences of social reformers, the congresses for international statistics and for social science bear witness of its force. Everywhere we hear of the development of the constitution, of public law, of public opinion, of institutions, of forms of society, of theories of history. In a word, whatever views of history may be inculcated on the Universities by novelists or epigrammatists, it is certain that the best intellects and spirits of our day are labouring to see more of that invariable order, and of that principle of growth in the life of human societies and of the great society of mankind which nearly all men, more or less, acknowledge, and partially and unconsciously confirm."

This passage expresses admirably, I think, the tendencies of modern thought for good and evil.

For good. For surely it is good, and a thing to thank God for, that men should be more and more expecting order, searching for order, welcoming order. But for evil also. For young sciences, like young men, have their time of wonder, hope, imagination, and of passion too, and haste, and bigotry. Dazzled, and that pardonably, by the beauty of the few laws they may have discovered, they are too apt to erect them into gods, and to explain by them all matters in heaven and earth; and apt, too, as I think this author does, to patch them where they are weakest, by that most dangerous succedaneum of vague and grand epithets, which very often contain, each of them, an assumption far more important than the law to which they are tacked.

Such surely are the words which so often
occur in this passage—"Invariable, continual,
immutable, inevitable, irresistible." There is
an ambiguity in these words, which may lead—
which I believe does lead—to most unphilosophical conclusions. They are used very much
as synonyms; not merely in this passage, but
in the mouths of men. Are you aware that
those who carelessly do so, blink the whole of
the world-old argument between necessity and
free-will? Whatever may be the rights of that
quarrel, they are certainly not to be assumed
in a passing epithet. But what else does the
writer do, who tells us that an inevitable
sequence, an irresistible growth, exists in the
moral as well as in the physical world; and then
says, as a seemingly identical statement, that it
is the crown of philosophy to see immutable
law, even in the complex action of human life?

The crown of philosophy? Doubtless it
is so. But not a crown, I should have thought,
which has been reserved as the special glory

of these latter days. Very early, at least in the known history of mankind, did Philosophy (under the humble names of Religion and Common Sense) see most immutable, and even eternal, laws, in the complex action of human life, even the laws of right and wrong; and called them The Everlasting Judgments of God, to which a confused and hardworked man was to look; and take comfort, for all would be well at last. By fair induction (as I believe) did man discover, more or less clearly, those eternal laws: by repeated verifications of them in every age, man has been rising, and will yet rise, to clearer insight into their essence, their limits, their practical results. And if it be these, the old laws of right and wrong, which this author and his school call invariable and immutable, we shall, I trust, most heartily agree with them; only wondering why a moral government of the world seems to them so very recent a discovery.

But we shall not agree with them, I trust,
when they represent these invariable and immutable laws as resulting in any inevitable
sequence, or irresistible growth. We shall not
deny a sequence—Reason forbids that; or
again, a growth—Experience forbids that: but
we shall be puzzled to see why a law, because
it is immutable itself, should produce inevitable results; and if they quote the facts of
material nature against us, we shall be ready
to meet them on that very ground, and ask:
—You say that as the laws of matter are
inevitable, so probably are the laws of human
life? Be it so: but in what sense are the laws
of matter inevitable? Potentially, or actually?
Even in the seemingly most uniform and
universal law, where do we find the inevitable or the irresistible? Is there not in nature
a perpetual competition of law against law,
force against force, producing the most end-
less and unexpected variety of results? Cannot each law be interfered with at any

moment by some other law, so that the first law, though it may struggle for the mastery, shall be for an indefinite time utterly defeated? The law of gravity is immutable enough : but do all stones inevitably fall to the ground? Certainly not, if I choose to catch one, and keep it in my hand. It remains there by laws; and the law of gravity is there too, making it feel heavy in my hand: but it has not fallen to the ground, and will not, till I let it. So much for the inevitable action of the laws of gravity, as of others. Potentially, it is immutable; but actually, it can be conquered by other laws.

I really beg your pardon for occupying you here with such truisms: but I must put the students of this University in mind of them, as long as too many modern thinkers shall choose to ignore them.

Even if then, as it seems to me, the history of mankind depended merely on physical laws, analogous to those which govern the rest of nature, it would be a hopeless task for us to discover an inevitable sequence in History, even though we might suppose that such existed. But as long as man has the mysterious power of breaking the laws of his own being, such a sequence not only cannot be discovered, but it cannot exist. For man can break the laws of his own being, whether physical, intellectual, or moral. He breaks them every day, and has always been breaking them. The greater number of them he cannot obey till he knows them. And too many of them he cannot know, alas, till he has broken them; and paid the penalty of his ignorance. He does not, like the brute or the vegetable, thrive by laws of which he is not conscious: but by laws of which he becomes gradually conscious; and which he can disobey after all. And therefore it seems to me very like a juggle of words to draw analogies from the physical and irrational world, and apply them to the moral and rational world; and most unwise to bridge over the gulf between the two by such adjectives as "irresistible" or "inevitable," such nouns as "order, sequence, law"—which must bear an utterly different meaning, according as they are applied to physical beings or to moral ones.

Indeed, so patent is the ambiguity, that I cannot fancy that it has escaped the author and his school; and am driven, by mere respect for their logical powers, to suppose that they mean no ambiguity at all; that they do not conceive of irrational beings as differing from rational beings, or the physical from the moral, or the body of man from his spirit, in kind and property; and that the immutable laws which they represent as governing human life and history have nothing at all to do with those laws of right and wrong, which I intend to set forth to you, as the "everlasting judgments of God."

In which case, I fear, they must go their way, and we ours, confessing that there is

an order, and there is a law, for man; and
that if he disturb that order, or break that
law in anywise, they will prove themselves
too strong for him, and reassert themselves,
and go forward, grinding him to powder, if
he stubbornly try to stop their way. But
asserting too, that his disobedience to them,
even for a moment, has disturbed the natural
course of events, and broken that inevitable
sequence, which we may find indeed, in our
own imaginations, as long as we sit with
a book in our studies : but which vanishes
the moment that we step outside into practical contact with life; and, instead of talking
cheerfully of a necessary and orderly progress,
find ourselves more inclined to cry with the
cynical man of the world:

"All the windy ways of men,
    Are but dust that rises up ;
And is lightly laid again."

The usual rejoinder to this argument is
to fall back upon man's weakness and igno-

rance, and to take refuge in the infinite unknown. Man, it is said, may of course interfere a little with some of the less important laws of his being : but who is he, to grapple with the more vast and remote ones ? Because he can prevent a pebble from falling, is he to suppose that he can alter the destiny of nations, and grapple forsooth with "the eternities and the immensities," and so forth ? The argument is very powerful : but addrest rather to the imagination than the reason. It is, after all, another form of the old omne ignotum pro magnifico; and we may answer, I think fairly—About the eternities and immensities we know nothing, not having been there as yet; but it is a mere assumption to suppose, without proof, that the more remote and impalpable laws are more vast, in the sense of being more powerful (the only sense which really bears upon the argument), than the laws which are palpably at work around us all day long; and

if we are capable of interfering with almost
every law of human life which we know of
already, it is more philosophical to believe
(till disproved by actual failure) that we can
interfere with those laws of our life which
we may know hereafter. Whether it will
pay us to interfere with them, is a different question. It is not prudent to interfere
with the laws of health, and it may not be
with other laws, hereafter to be discovered.
I am only pleading that man can disobey
the laws of his´ being; that such power
has always been a disturbing force in the
progress of the human race, which modern
theories too hastily overlook; and that the
science of history (unless the existence of the
human will be denied) must belong rather to
the moral sciences, than to that "positive
science" which seems to me inclined to reduce
all human phænomena under physical laws,
hastily assumed, by the old fallacy of μετα-
βάσις εἰς ἄλλο γένος, to apply where there is no

proof whatsoever that they do or even can apply.

As for the question of the existence of the human will—I am not here, I hope, to argue that. I shall only beg leave to assume its existence, for practical purposes. I may be told (though I trust not in this University), that it is, like the undulatory theory of light, an unphilosophical "hypothesis." Be that as it may, it is very convenient (and may be for a few centuries to come) to retain the said "hypothesis," as one retains the undulatory theory; and for the simple reason, that with it one can explain the phænomena tolerably; and without it cannot explain them at all.

A dread (half-unconscious, it may be) of this last practical result, seems to have crossed the mind of the author on whom I have been commenting; for he confesses, honestly enough (and he writes throughout like an honest man) that in human life "no rational

thinker hopes to discover more than some
few primary actions of law, and some approximative theory of growth." I have higher
hopes of a possible science of history; because
I fall back on those old moral laws, which I
think he wishes to ignore: but I can conceive
that he will not; because he cannot, on his
own definitions of law and growth. They are
(if I understand him aright) to be irresistible
and inevitable. I say that they are not so,
even in the case of trees and stones; much
more in the world of man. Facts, when he
goes on to verify his theories, will leave him
with a very few primary actions of law, a very
faint approximative theory; because his theories, in plain English, will not work. At the
first step, at every step, they are stopped
short by those disturbing forces, or at least
disturbed phænomena, which have been as
yet, and probably will be hereafter, attributed
(as the only explanation of them) to the
existence, for good and evil, of a human will.

Let us look in detail at a few of these disturbances of anything like inevitable or irresistible movement. Shall we not, at the very first glance, confess—I am afraid only too soon—that there always have been fools therein; fools of whom no man could guess, or can yet, what they were going to do next or why they were going to do it? And how, pray, can we talk of the inevitable, in the face of that one miserable fact of human folly, whether of ignorance or of passion, folly still? There may be laws of folly, as there are laws of disease; and whether there are or not, we may learn much wisdom from folly; we may see what the true laws of humanity are, by seeing the penalties which come from breaking them: but as for laws which work of themselves, by an irresistible movement,—how can we discover such in a past in which every law which we know has been outraged again and again? Take one of the highest instances—the progress

of the human intellect—I do not mean just
now the spread of conscious science, but
of that unconscious science which we call
common sense. What hope have we of lay-
ing down exact laws for its growth, in a
world wherein it has been ignored, insulted,
crushed, a thousand times, sometimes in whole
nations and for whole generations, by the
stupidity, tyranny, greed, caprice of a single
ruler; or if not so, yet by the mere superstition, laziness, sensuality, anarchy of the mob?
How, again, are we to arrive at any exact
laws of the increase of population, in a race
which has had, from the beginning, the abnormal and truly monstrous habit of slaugh-
tering each other, not for food—for in a race
of normal cannibals, the ratio of increase or
decrease might easily be calculated—but use-
lessly, from rage, hate, fanaticism, or even
mere wantonness? No man is less inclined
than I to undervalue vital statistics, and their
already admirable results: but how can they

help us, and how can we help them, in looking
at such a past as that of three-fourths of the
nations of the world? Look—as a single instance among too many—at that most noble
nation of Germany, swept and stunned, by
peasant wars, thirty years' wars, French
wars, and after each hurricane, blossoming
up again into brave industry and brave
thought, to be in its turn cut off by a fresh
storm ere it could bear full fruit: doing nevertheless such work, against such fearful dis-
advantages, as nation never did before; and
proving thereby what she might have done
for humanity, had not she, the mother of all
European life, been devoured, generation after
generation, by her own unnatural children.
Nevertheless, she is their mother still; and
her history, as I believe, the root-history of
Europe: but it is hard to read—the sibylline
leaves are so fantastically torn, the characters
so blotted out by tears and blood.

And if such be the history of not one

nation only, but of the average, how, I ask, are we to make calculations about such a species as man? Many modern men of science wish to draw the normal laws of human life from the average of humanity: I question whether they can do so; because I do not believe the average man to be the normal man, exhibiting the normal laws: but a very abnormal man, diseased and crippled, but even if their method were correct, it could work in practice, only if the destinies of men were always decided by majorities: and granting that the majority of men have common sense, are the minority of fools to count for nothing? Are they powerless? Have they had no influence on History? Have they even been always a minority, and not at times a terrible majority, doing each that which was right in the sight of his own eyes? You can surely answer that question for yourselves. As far as my small knowledge of History goes, I think it may be

proved from facts, that any given people, down
to the lowest savages, has, at any period of
its life, known far more than it has done;
known quite enough to have enabled it to
have got on comfortably, thriven, and developed; if it had only done, what no man does, all
that it knew it ought to do, and could do. St
Paul's experience of himself is true of all
mankind—"The good which I would, I do not;
and the evil which I would not, that I do."
The discrepancy between the amount of knowledge and the amount of work, is one of the
most patent and most painful facts which
strikes us in the history of man; and one not
certainly to be explained on any theory of
man's progress being the effect of inevitable
laws, or one which gives us much hope of
ascertaining fixed laws for that progress.

And bear in mind, that fools are not
always merely imbecile and obstructive; they
are at times ferocious, dangerous, mad. There
is in human nature what Göthe used to call

3

a demoniac element, defying all law, and all induction; and we can, I fear, from that one cause, as easily calculate the progress of the human race, as we can calculate that of the vines upon the slopes of Ætna, with the lava ready to boil up and overwhelm them at any and every moment. Let us learn, in God's name, all we can, from the short intervals of average peace and common sense: let us, or rather our grandchildren, get precious lessons from them for the next period of sanity. But let us not be surprised, much less disheartened, if after learning a very little, some unexpected and truly demoniac factor, Anabaptist war, French revolution, or other, should toss all our calculations to the winds, and set us to begin afresh, sadder and wiser men. We may learn, doubtless, even more of the real facts of human nature, the real laws of human history, from these critical periods, when the root-fibres of the human heart are laid bare, for good and evil, than from any smooth and respectable periods of peace and plenty: nevertheless their lessons are not statistical, but moral.

But if human folly has been a disturbing force for evil, surely human reason has been a disturbing force for good. Man can not only disobey the laws of his being, he can also choose between them, to an extent which science widens every day, and so become, what he was meant to be, an artificial being; artificial in his manufactures, habits, society, polity—what not? All day long he has a free choice between even physical laws, which mere things have not, and which make the laws of mere things inapplicable to him. Take the simplest case. If he falls into the water, he has his choice whether he will obey the laws of gravity and sink, or by other laws perform the (to him) artificial process of swimming, and get ashore. True, both would happen by law: but he has his choice which law shall conquer, sink or swim. We have

yet to learn why whole nations, why all mankind may not use the same prudential power
as to which law they shall obey,—which, without breaking it, they shall conquer and re-
press, as long as seems good to them.

It is true, nature must be obeyed in order
that she may be conquered: but then she is
to be CONQUERED. It has been too much the
fashion of late to travestie that great dictum
of Bacon's into a very different one, and say,
Nature must be obeyed because she cannot
be conquered; thus proclaiming the impotence
of science to discover anything save her own
impotence—a result as contrary to fact, as to
Bacon's own hopes of what science would do for
the welfare of the human race. For what is all
human invention, but the transcending and con-
quering one natural law by another? What is
the practical answer which all mankind has
been making to nature and her pretensions,
whenever it has progressed one step since
the foundation of the world: by which all dis-

coverers have discovered, all teachers taught:
by which all polities, kingdoms, civilizations,
arts, manufactures, have established themselves; all who have raised themselves above
the mob have faced the mob, and conquered the mob, crucified by them first and
worshipped by them afterwards : by which
the first savage conquered the natural law
which put wild beasts in the forest, by killing
them; conquered the natural law which makes
raw meat wholesome, by cooking it; conquered the natural law which made weeds
grow at his hut door, by rooting them up,
and planting corn instead; and won his
first spurs in the great battle of man against
nature, proving thereby that he was a man,
and not an ape? What but this?—"Nature is strong, but I am stronger. I know
her worth, but I know my own. I trust her
and her laws, but my trusty servant she shall
be, and not my tyrant; and if she interfere
with my ideal, even with my personal com-

fort, then Nature and I will fight it out to the
last gasp, and Heaven defend the right!"

In forgetting this, in my humble opinion,
lay the error of the early, or *laissez faire*
School of Political Economy. It was too
much inclined to say to men : "You are the
puppets of certain natural laws. Your own
free-will and choice, if they really exist, exist
merely as a dangerous disease. All you can do
is to submit to the laws, and drift whither-
soever they may carry you, for good or evil."
But not less certainly was the same blame to be
attached to the French Socialist School. It,
though based on a revolt from the Philosophie du neant, philosophie de la misère, as it
used to term the laissez faire School, yet re-
tained the worst fallacy of its foe, namely, that
man was the creature of circumstances ; and
denied him just as much as its antagonist the
possession of freewill, or at least the right to
use freewill on any large scale.

The *laissez faire* School was certainly

the more logical of the two. With them, if man was the creature of circumstances, those circumstances were at least defined for him by external laws which he had not created: while the Socialists, with Fourier at their head (as it has always seemed to me), fell into the extraordinary paradox of supposing that though man was the creature of circumstances, he was to become happy by creating the very circumstances which were afterwards to create him. But both of them erred, surely, in ignoring that self-arbitrating power of man, by which he can, for good or for evil, rebel against and conquer circumstance.

I am not, surely, overstepping my province as Professor of History, in alluding to this subject. Just notions of Political Economy are absolutely necessary to just notions of History; and I should wish those young gentlemen who may attend my Lectures, to go first, were it possible, to my more learned brother, the Professor of Political Economy,

and get from him not merely exact habits of
thought, but a knowledge which I cannot give,
and yet which they ought to possess. For to
take the very lowest ground, the first fact
of history is, Bouche va toujours; whatever
men have or have not done, they have always
eaten, or tried to eat; and the laws which
regulate the supply of the first necessaries of
life are, after all, the first which should be
learnt, and the last which should be ignored.

The more modern school, however, of Political Economy while giving due weight to
circumstance, has refused to acknowledge it
as the force which ought to determine all
human life; and our greatest living political
economist has, in his Essay on Liberty, put
in a plea unequalled since the Areopagitica of
Milton, for the self-determining power of the
individual, and for his right to use that power.

But my business is not with rights, so
much as with facts; and as a fact, surely, one
may say, that this inventive reason of man

has been, in all ages, interfering with any thing like an inevitable sequence or orderly progress of humanity. Some of those writers, indeed, who are most anxious to discover an exact order, are most loud in their complaints that it has been interfered with by over-legislation ; and rejoice that mankind is returning to a healthier frame of mind, and leaving nature alone to her own work in her own way. I do not altogether agree with their complaints ; but of that I hope to speak in subsequent lectures. Meanwhile, I must ask, if (as is said) most good legislation now-a-days consists in repealing old laws which ought never to have been passed ; if (as is said) the great fault of our forefathers was that they were continually setting things wrong, by intermeddling in matters political, economic, religious, which should have been let alone, to develop themselves in their own way, what becomes of the inevitable laws, and the continuous progress, of the human mind ?

Look again at the disturbing power, not merely of the general reason of the many, but of the genius of the few. I am not sure, but that the one fact, that genius is occasionally present in the world, is not enough to prevent our ever discovering any regular sequence in human progress, past or future.

Let me explain myself. In addition to the infinite variety of individual characters continually born (in itself a cause of perpetual disturbance), man alone of all species has the faculty of producing, from time to time, individuals immeasurably superior to the average in some point or other, whom we call men of genius. Like Mr Babbage's calculating machine, human nature gives millions of orderly respectable common-place results, which any statistician can classify, and enables hasty philosophers to say—It always has gone on thus; it must go on thus always; when behold, after many millions of orderly results, there turns up a seemingly disorderly, a certainly

unexpected, result, and the law seems broken
(being really superseded by some deeper law)
for that once, and perhaps never again for
centuries. Even so it is with man, and the
physiological laws which determine the earthly
appearance of men. Laws there are, doubt
it not; but they are beyond us : and let our
induction be as wide as it may, they will
baffle it; and great nature, just as we fancy
we have found out her secret, will smile in
our faces as she brings into the world a man,
the like of whom we have never seen, and
cannot explain, define, classify—in one word,
a genius. Such do, as a fact, become leaders
of men into quite new and unexpected paths,
and for good or evil, leave their stamp upon
whole generations and races. Notorious as
this may be, it is just, I think, what most modern theories of human progress ignore. They
take the actions and the tendencies of the
average many, and from them construct their
scheme : a method not perhaps quite safe

were they dealing with plants or animals; but
what if it be the very peculiarity of this
fantastic and altogether unique creature called
man, not only that he develops, from time to
time, these exceptional individuals, but that
they are the most important individuals of
all? that his course is decided for him not
by the average many, but by the extraordinary few; that one Mahommed, one Luther,
one Bacon, one Napoleon, shall change the
thoughts and habits of millions?—So that instead of saying that the history of mankind
is the history of its masses, it would be much
more true to say, that the history of mankind
" is the history of its great men; and that a
true philosophy of history ought to declare the
laws—call them physical, spiritual, biological,
or what we choose—by which great minds
have been produced into the world, as necessary results, each in his place and time.

That would be a science indeed; how far
we are as yet from any such, you know as

well as I. As yet, the appearance of great
minds is as inexplicable to us as if they had
dropped among us from another planet. Who
will tell us why they have arisen when
they did, and why they did what they did,
and nothing else? I do not deny that such
a science is conceivable; because each mind,
however great or strange, may be the result
of fixed and unerring laws of life: and it is
conceivable, too, that such a science may so
perfectly explain the past, as to be able to
predict the future; and tell men when a fresh
genius is likely to arise and of what form
his intellect will be. Conceivable: but I fear
only conceivable; if for no other reason, at
least for this one. We may grant safely that
the mind of Luther was the necessary result
of a combination of natural laws. We may
go further, and grant, but by no means safely,
that Luther was the creature of circumstances,
that there was no self-moving originality in
him, but that his age made him what he was.

To some modern minds these concessions re-
move all difficulty and mystery: but not, I
trust, to our minds. For does not the very
puzzle de quo agitur remain equally real;
namely, why the average of Augustine monks,
the average of German men, did not, by
being exposed to the same average circumstances as Luther, become what Luther
was? But whether we allow Luther to have
been a person with an originally different
character from all others, or whether we hold
him to have been the mere puppet of outside
influences, the first step towards discovering
how he became what he was, will be to find
out what he was.   It will be more easy, and,
I am sorry to say, more common to settle
beforehand our theory, and explain by it
such parts of Luther as will fit it; and call
those which will not fit it hard names.
History is often so taught, and the method
is popular and lucrative. But we here shall
be of opinion, I am sure, that we can only

learn causes through their effects; we can only
learn the laws which produced Luther, by
learning Luther himself; by analyzing his
whole character; by gauging all his powers;
and that—unless the less can comprehend
the greater—we cannot do till we are more
than Luther himself. I repeat it. None can
comprehend a man, unless he be greater than
that man. He must be not merely equal
to him, because none can see in another
elements of character which he has not already seen in himself: he must be greater;
because to comprehend him thoroughly, he
must be able to judge the man's failings as
well as his excellencies; to see not only why
he did what he did, but why he did not do
more: in a word, he must be nearer than his
object is to the ideal man.

And if it be assumed that I am quibbling
on the words "comprehend" and "greater,"
that the observer need be greater only potentially, and not in act; that all the compre-

hension required of him, is to have in himself
the germs of other men's faculties, without
having developed those germs in life; I must
still stand to my assertion. For such a rejoinder ignores the most mysterious element
of all character, which we call strength: by
virtue of which, of two seemingly similar
characters, while one does nothing, the other
shall do great things; while in one the germs
of intellect and virtue remain comparatively
embryotic, passive, and weak, in the other
these same germs shall develop into manhood,
action, success. And in what that same
strength consists, not even the dramatic
imagination of a Shakspeare could discover.
What are those heart-rending sonnets of his,
but the confession that over and above all
his powers he lacked one thing, and knew
not what it was, or where to find it—and
that was—to be strong?

And yet he who will give us a science
of great men, must begin by having a larger

heart, a keener insight, a more varying human experience, than Shakespeare's own; while those who offer us a science of little men, and attempt to explain history and progress by laws drawn from the average of mankind, are utterly at sea the moment they come in contact with the very men whose actions make the history, to whose thought the progress is due. And why? Because (so at least I think) the new science of little men can be no science at all: because the average man is not the normal man, and never yet has been; because the great man is rather the normal man, as approaching more nearly than his fellows to the true "norma" and standard of a complete human character; and therefore to pass him by as a mere irregular sport of nature, an accidental giant with six fingers and six toes, and to turn to the mob for your theory of humanity, is (I think) about as wise as to ignore the Apollo and the Theseus, and to determine the pro-

portions of the human figure from a crowd of
dwarfs and cripples.

No, let us not weary ourselves with narrow
theories, with hasty inductions, which will, a
century hence, furnish mere matter for a
smile. Let us confine ourselves, at least in the
present infantile state of the anthropologic
sciences, to facts; to ascertaining honestly and
patiently the thing which has been done;
trusting that if we make ourselves masters of
them, some rays of inductive light will be
vouchsafed to us from Him who truly com-
prehends mankind, and knows what is in
man, because He is the Son of Man; who
has His own true theory of human progress,
His own sound method of educating the
human race, perfectly good, and perfectly wise,
and at last, perfectly victorious; which nevertheless, were it revealed to us to-morrow, we
could not understand; for if he who would comprehend Luther must be more than Luther,
what must he be, who would comprehend God?

Look again, as a result of the disturbing force of genius, at the effects of great inventions—how unexpected, complex, subtle, all but miraculous—throwing out alike the path of human history, and the calculations of the student. If physical discoveries produced only physical or economic results—if the invention of printing had only produced more books, and more knowledge—if the invention of gunpowder had only caused more or less men to be killed—if the invention of the spinning-jenny had only produced more cotton-stuffs, more employment, and therefore more human beings,—then their effects would have been, however complex, more or less subjects of exact computation.

But so strangely interwoven is the physical and spiritual history of man, that material inventions produce continually the most unexpected spiritual results. Printing becomes a religious agent, causes not merely more books, but a Protestant Reformation; then

again, through the Jesuit literature, helps to
a Romanist counter-reformation; and by the
clashing of the two, is one of the great causes
of the Thirty years war, one of the most dis-
astrous checks which European progress ever
suffered. Gunpowder, again, not content with
killing men, becomes unexpectedly a political
agent; "the villanous saltpetre," as Ariosto
and Shakspeare's fop complain, "does to death
many a goodly gentleman," and enables the
masses to cope, for the first time, with knights
in armour; thus forming a most important
agent in the rise of the middle classes; while
the spinning-jenny, not content with furnishing facts for the political economist, and employment for millions, helps to extend slavery
in the United States, and gives rise to moral
and political questions, which may have, ere
they be solved, the most painful conse-
quences to one of the greatest nations on
earth.

So far removed is the sequence of human

history from any thing which we can call irresistible or inevitable. Did one dare to deal
in epithets, crooked, wayward, mysterious, in-
calculable, would be those which would rather
suggest themselves to a man looking steadily
not at a few facts here and there, and not
again at some hasty bird's-eye sketch, which
he chooses to call a whole: but at the actual
whole, fact by fact, step by step, and alas!
failure by failure, and crime by crime.

Understand me, I beg. I do not wish
(Heaven forbid!) to discourage inductive
thought; I do not wish to undervalue exact
science. I only ask that the moral world,
which is just as much the domain of inductive science as the physical one, be not
ignored; that the tremendous difficulties of
analyzing its phenomena be fairly faced; and
the hope given up, at least for the present, of
forming any exact science of history; and I
wish to warn you off from the too common
mistake of trying to explain the mysteries of

the spiritual world by a few roughly defined
physical laws (for too much of our modern
thought does little more than that); and of
ignoring as old fashioned, or even super-
stitious, those great moral laws of history,
which are sanctioned by the experience of ages.

Foremost among them stands a law which
I must insist on, boldly and perpetually, if I
wish (as I do wish) to follow in the footsteps
of Sir James Stephen : a law which man has
been trying in all ages, as now, to deny, or
at least to ignore; though he might have
seen it if he had willed, working steadily in all
times and nations.   And that is—that as the
fruit of righteousness is wealth and peace,
strength and honour; the fruit of unrighteous-
ness is poverty and anarchy, weakness and
shame.   It is an ancient doctrine, and yet
one ever young.   The Hebrew prophets
preached it long ago, in words which are
fulfilling themselves around us every day,
and which no new discoveries of science will

abrogate, because they express the great root-law, which disobeyed, science itself cannot get a hearing.

For not upon mind, gentlemen, not upon mind, but upon morals, is human welfare founded. The true subjective history of man is the history not of his thought, but of his conscience; the true objective history of man is not that of his inventions, but of his vices and his virtues. So far from morals depending upon thought, thought, I believe, depends on morals. In proportion as a nation is righteous,—in proportion as common justice is done between man and man, will thought grow rapidly, securely, triumphantly; will its discoveries be cheerfully accepted, and faithfully obeyed, to the welfare of the whole commonweal. But where a nation is corrupt, that is, where the majority of individuals in it are bad, and justice is not done between man and man, there thought will wither, and science will be either crushed by frivolity and

sensuality, or abused to the ends of tyranny,
ambition, profligacy, till she herself perishes,
amid the general ruin of all good things; as
she has done in Greece, 'in Rome, in Spain,
in China, and many other lands. Laws of
economy, of polity, of health, of all which
makes human life endurable, may be ignored
and trampled under foot, and are too often,
every day, for the sake of present greed, of
present passion; self-interest may become, and
will become, more and more blinded, just
in proportion as it is not enlightened by
virtue; till a nation may arrive, though,
thank God, but seldom, at that state of frantic
recklessness which Salvian describes among
his Roman countrymen in Gaul, when, while
the Franks were thundering at their gates,
and starved and half-burnt corpses lay about
the unguarded streets, the remnant, like that
in doomed Jerusalem of old, were drinking,
dicing, ravishing, robbing the orphan and the
widow, swindling the poor man out of his

plot of ground, and sending meanwhile to
the tottering Cæsar at Rome, to ask, not for
armies, but for Circensian games.

We cannot see how science could have
bettered those poor Gauls. And we can conceive, surely, a nation falling into the same
madness, and crying, " Let us eat and drink,
for to-morrow we die," in the midst of rail-
roads, spinning-jennies, electric telegraphs,
and crystal palaces, with infinite blue-books
and scientific treatises ready to prove to
them, what they knew perfectly well already,
that they were making a very unprofitable
investment, both of money and of time.

For science indeed is great: but she is
not the greatest. She is an instrument, and
not a power ; beneficent or deadly, according
as she is wielded by the hand of virtue or
of vice. But her lawful mistress, the only
one which can use her aright, the only one
under whom she can truly grow, and prosper,
and prove her divine descent, is Virtue, the

likeness of Almighty God. This, indeed, the Hebrew Prophets, who knew no science in one sense of the word, do not expressly say: but it is a corollary from their doctrine, which we may discover for ourselves, if we will look at the nations round us now, if we will look at all the nations which have been. Even Voltaire himself acknowledged that; and when he pointed to the Chinese as the most prosperous nation upon earth, ascribed their prosperity uniformly to their virtue. We now know that he was wrong in fact : for we have discovered that Chinese civilization is one not of peace and plenty, but of anarchy and wretchedness. But that fact only goes to corroborate the belief, which (strange juxtaposition!) was common to Voltaire and the old Hebrew Prophets at whom he scoffed, namely, that virtue is wealth, and vice is ruin. For we have found that these Chinese, the ruling classes of them at least, are an especially unrighteous people; rotting upon

the rotting remnants of the wisdom and virtue of their forefathers, which now lives only
on their lips in flowery maxims about justice
and mercy and truth, as a cloak for practical
hypocrisy and villany ; and we have discovered also, as a patent fact, just what the
Hebrew Prophets would have foretold us—
that the miseries and horrors which are now
destroying the Chinese Empire, are the direct
and organic results of the moral profligacy
of its inhabitants.

I know no modern nation, moreover, which
illustrates so forcibly as China the great historic law which the Hebrew Prophets pro-
claim ; and that is this:—That as the prosperity of a nation is the correlative of their
morals, so are their morals the correlative of
their theology. As a people behaves, so it
thrives ; as it believes, so it behaves. Such
as his Gods are, such will the man be; down
to that lowest point which too many of the
Chinese seem to have reached, where, having

no Gods, he himself becomes no man; but
(as I hear you see him at the Australian
diggings) abhorred for his foul crimes even
by the scum of Europe.

I do not say that the theology always produces the morals, any more than that the
morals always produces the theology. Each
is, I think, alternately cause and effect. Men
make the Gods in their own likeness; then
they copy the likeness they have set up. But
whichever be cause, and whichever effect, the
law, I believe, stands true, that on the two
together depend the physical welfare of a
people. History gives us many examples, in
which superstition, many again in which profligacy, have been the patent cause of a nation's degradation. It does not, as far as I
am aware, give us a single case of a nation's
thriving and developing when deeply infected
with either of those two vices.

These, the broad and simple laws of moral
retribution, we may see in history; and

(I hope) something more than them; something of a general method, something of an upward progress, though any thing but an irresistible or inevitable one. For I have not argued that there is no order, no progress—God forbid. Were there no order to be found, what could the student with a man's reason in him do, but in due time go mad?—Were there no progress, what could the student with a man's heart within him do, but in due time break his heart, over the sight of a chaos of folly and misery irredeemable?— I only argue that the order and the progress of human history cannot be similar to those which govern irrational beings, and cannot (without extreme danger) be described by metaphors (for they are nothing stronger) drawn from physical science. If there be an order, a progress, they must be moral; fit for the guidance of moral beings; limited by the obedience which those moral beings pay to what they know.

And such an order, such a progress as that, I have good hope that we shall find in history.

We shall find, as I believe, in all the ages, God educating man; protecting him till he can go alone, furnishing him with the primary necessaries, teaching him, guiding him, inspiring him, as we should do to our children; bearing with him, and forgiving him too, again and again, as we should do: but teaching him withal (as we shall do if we be wise) in great part by his own experience, making him test for himself, even by failure and pain, the truth of the laws which have been given him; discover for himself, as much as possible, fresh laws, or fresh applications of laws; and exercising his will and faculties, by trusting him to himself wherever he can be trusted without his final destruction. This is my conception of history, especially of Modern History—of history since the Revelation of our Lord Jesus Christ. I express myself feebly

enough, I know. And even could I express
what I mean perfectly, it would still be but a
partial analogy, not to be pushed into details.
As I said just now, were the true law of
human progress revealed to us to-morrow, we
could not understand it.

For suppose that the theory were true,
which Dr Temple of Rugby has lately put .
into such noble words : suppose that, as he
says, "The power whereby the present ever
gathers into itself the results of the past,
transforms the human race into a colossal man,
whose life reaches from the creation to the
day of judgment. The successive generations
of men, are days in this man's life. The discoveries and inventions which characterize the
different epochs of the world, are this man's
works. The creeds and doctrines, the opinions
and principles of the successive ages, are his
thoughts. The state of society at different
times, are his manners. He grows in knowledge, in self-control, in visible size, just as we

do." Suppose all this; and suppose too, that
God is educating this his colossal child, as we
educate our own children ; it will hardly follow from thence that his education would be,
as Dr Temple says it is, precisely similar to
ours.

Analogous it may be, but not precisely
similar ; and for this reason : That the collec-
tive man, in the theory, must be infinitely
more complex in his organization than the
individuals of which he is composed. While
between the educator of the one and of the
other, there is simply the difference between
a man and God. How much more complex
then must his education be ! how all-inscru-
table to human minds much in it !—often
as inscrutable as would our training of our
children seem to the bird brooding over her
young ones in the nest. The parental relations in all three cases may be—the Scriptures
say that they are—expansions of the same
great law ; the key to all history may be con-

tained in those great words—"How often would I have gathered thy children as a hen gathereth her chickens under her wings." Yet even there, the analogy stops short—"but thou wouldest not" expresses a new element, which has no place in the training of the nestling by the dam, though it has place in our training of our children ; even that self-will, that power of disobedience, which is the dark side of man's prerogative as a rational and self-cultivating being. Here that analogy fails, as we should have expected it to do; and in a hundred other points it fails, or rather transcends so utterly its original type, that mankind seems, at moments, the mere puppet of those laws of natural selection, and competition of species, of which we have heard so much of late ; and, to give a single instance, the seeming waste, of human thought, of human agony, of human power, seems but another instance of that inscrutable prodigality of nature, by which, of a thousand acorns

5

dropping to the ground, but one shall become
the thing it can become, and grow into a
builder oak, the rest be craunched up by the
nearest swine.

Yet these dark passages of human life
may be only necessary elements of the complex education of our race; and as much
mercy under a fearful shape, as ours when
we put the child we love under the surgeon's
knife. At least we may believe so; believe
that they have a moral end, though that end
be unseen by us; and without any rash or
narrow prying into final causes (a trick as
fatal to historic research as Bacon said it was
to science), we may justify God by faith,
where we cannot justify Him by experience.

Surely this will be the philosophic method.
If we seem to ourselves to have discovered
a law, we do not throw it away the moment
we find phænomena which will not be explained by it. We use those phænomena to
correct and to expand our law. And this

belief that History is "God educating man," is no mere hypothesis; it results from the observation of thousands of minds, throughout thousands of years. It has long seemed —I trust it will seem still—the best explanation of the strange deeds of that strange being man: and where we find in history facts which seem to contradict it, we shall not cast away rashly or angrily either it or them : but if we be Bacon's true disciples, we shall use them patiently and reverently to correct and expand our notions of the law itself, and rise thereby to more deep and just conceptions of education, of man, and—it may be—of God Himself.

In proportion as we look at history thus; searching for effective, rather than final causes, and content to see God working everywhere, without impertinently demanding of Him a reason for His deeds, we shall study in a frame of mind equally removed from superstition on the one hand, and necessitarianism

on the other. We shall not be afraid to confess natural agencies: but neither shall we be
afraid to confess those supernatural causes
which underlie all existence, save God's
alone.

We shall talk of more than of an over-
ruling Providence. That such exists, will
seem to us a patent fact. But it will seem to
us somewhat Manichæan to believe that the
world is ill made, mankind a failure, and that
all God has to do with them, is to set them
right here and there, when they go intole-
rably wrong. We shall believe not merely
in an over-ruling Providence, but (if I may
dare to coin a word) in an under-ruling one,
which has fixed for mankind eternal laws of
life, health, growth, both physical and spiritual; in an around-ruling Providence, like-
wise, by which circumstances, that which
stands around a man, are perpetually arranged, it may be, are fore-ordained, so that each
law shall have at least an opportunity of

taking effect on the right person, in the right time and place; and in an in-ruling Providence, too, from whose inspiration comes all true thought, all right feeling; from whom, we must believe, man alone of all living things known to us inherits that mysterious faculty of perceiving the law beneath the phenomena, by virtue of which, he is a *man*.

But we can hold all this, surely, and equally hold all which natural science may teach us. Hold what natural science teaches? We shall not dare not to hold it. It will be sacred in our eyes. All light which science, political, economic, physiological, or other, can throw upon the past, will be welcomed by us, as coming from the Author of all light. To ignore it, even to receive it suspiciously and grudgingly, we shall feel to be a sin against Him. We shall dread no "inroads of materialism;" because we shall be standing upon that spiritual ground which underlies— ay, causes — the material. All discoveries

of science, whether political or economic, whether laws of health or laws of climate, will
be accepted trustfully and cheerfully. And
when we meet with such startling speculations
as those on the influence of climate, soil, scenery
on national character, which have lately excited
so much controversy, we shall welcome them
at first sight, just because they give us hope
of order where we had seen only disorder,
law where we fancied chance: we shall verify
them patiently; correct them if they need
correction; and if proven, believe that they
have worked, and still work, οὐκ ἄνευ Θεοῦ,
as factors in the great method of Him who
has appointed to all nations their times, and
the bounds of their habitation, if haply they
might feel after Him, and find Him: though
He be not far from any one of them; for in
Him we live, and move, and have our being,
and are the offspring of God Himself.

I thus end what it seemed to me proper
to say in this, my Inaugural Lecture ; thank-

ing you much for the patience with which you
have heard me : and if I have in it too often
spoken of myself, and my own opinions, I can
only answer that it is a fault which has been
forced on me by my position, and which will
not occur again. It seemed to me that some
sort of statement of my belief was necessary,
if only from respect to a University from
which I have been long separated, and to
return to which is to me a high honour and a
deep pleasure ; and I cannot but be aware
(it is best to be honest) that there exists a
prejudice against me in the minds of better
men than I am, on account of certain early
writings of mine. That prejudice, I trust,
with God's help, I shall be able to dissipate.
At least whatever I shall fail in doing, this
University will find that I shall do one thing;
and that is, obey the Apostolic precept,
"Study to be quiet, and to do your own
business."

I have now to announce, that my lectures

will commence the first week in next February, and be spread over the Lent and Easter terms ; and that, meanwhile, if any Undergraduates wish to become members of my Class, I shall be most happy to see them at my own house, on Mondays, Wednesdays, or Fridays, at the hour of twelve, and tell them what books it seems to me they ought to read : always premising, that Gibbon, whether I may agree or disagree with him in details, will form the text-book on which they will be examined by me.

CAMBRIDGE: PRINTED AT THE UNIVERSITY PRESS.

SELECT LIST OF

𝔑𝔢𝔴 𝔚𝔬𝔯𝔨𝔰 𝔞𝔫𝔡 𝔑𝔢𝔴 𝔈𝔡𝔦𝔱𝔦𝔬𝔫𝔰

PUBLISHED BY

MACMILLAN AND CO.

CAMBRIDGE,

AND 23, HENRIETTA STREET, COVENT GARDEN, LONDON, W.C.

ONE SHILLING MONTHLY,

MACMILLAN'S
MAGAZINE.

EDITED BY DAVID MASSON.

VOLUMES I. AND II. ARE NOW READY,

Handsomely bound in extra cloth, price 7s. 6d. each.

AMONG THE CONTRIBUTORS TO THE VOLUMES ARE

THE AUTHOR OF "TOM BROWN'S SCHOOL DAYS."
THE AUTHOR OF "JOHN HALIFAX."

THE REV. F. D. MAURICE.	ALFRED TENNYSON.
R. MONCKTON MILNES, M.P.	PROFESSOR HUXLEY.
THE REV. J. W. BLAKESLEY.	G. S. VENABLES.
HENRY KINGSLEY.	PROFESSOR ANSTED.
F. LUSHINGTON.	J. M. LUDLOW.
ALEXANDER SMITH.	HERBERT COLERIDGE.
AURELIO SAFFI.	REV. J. LL. DAVIES.
&c.	*&c.*

23.11.60.
3,000 post.

A

POPULAR WORKS FOR THE YOUNG.

PRICE FIVE SHILLINGS EACH.

Tom Brown's School-Days.   By AN OLD BOY.
With a new Preface.   Seventh Edition.   Fcap. 8vo. 5s.

" *Those manly, honest thoughts, expressed in plain words, will, we trust, long
find an echo in thousands of English hearts.*"—QUARTERLY REVIEW.

Our Year.   A Child's Book in Prose and Rhyme.
By the Author of "John Halifax."   With numerous Illustrations
by CLARENCE DOBELL.   Royal 16mo. cloth, gilt leaves, 5s.

" *Just the book we could wish to see in the hands of every child ... written in
such an easy, chatty, kindly manner.*"—ENGLISH CHURCHMAN.

Mr. Kingsley's Heroes, or Greek Fairy Tales for my
Children.   New Edition, with Illustrations.
Royal 16mo. cloth, gilt leaves, 5s.

" *A welcome and delightful volume, for the stories are prose poems both as to
matter and manner.*"—ECLECTIC REVIEW.

Ruth and Her Friends.   A Story for Girls.
With Frontispiece.   Third Edition.
Royal 16mo, cloth, gilt leaves, 5s.

" *The tone is so thoroughly healthy, that we augur the happiest results from
its wide diffusion.*"—THE FREEMAN.

POPULAR WORKS FOR THE YOUNG—*Continued.*

Days of Old : Stories from Old English History for the Young.

By the Author of "RUTH AND HER FRIENDS." With Frontispiece. Royal 16mo. cloth, gilt leaves, 5s.

"*A delightful little book, full of interest and instruction . . . fine feeling, dramatic weight, and descriptive power in the stories.*"—LITERARY GAZETTE.

Agnes Hopetoun's Schools and Holidays : the Experience of a Little Girl.

By Mrs. OLIPHANT (Author of "Margaret Maitland"). With Frontispiece. Royal 16mo. cloth, gilt leaves, 5s.

"*One of Mrs. Oliphant's gentle, thoughtful stories. . . . described with exquisite reality . . . teaching the young pure and good lessons.*"—JOHN BULL.

Little Estella, and other Tales for the Young.

With Frontispiece. Royal 16mo. cloth, gilt leaves, 5s.

"*Very pretty, pure in conception, and simply, gracefully related . . . genuine story telling.*"—DAILY NEWS.

David, King of Israel. A History for the Young.

By J. WRIGHT. With Illustrations. Royal 16mo. cloth, gilt leaves, 5s.

"*An excellent book . . . well conceived, and well worked out.*"—LITERARY CHURCHMAN.

My First Journal : a Book for Children.

By GEORGIANA M. CRAIK, Author of "Lost and Won." With Frontispiece. Royal 16mo. cloth, gilt leaves. 4s. 6d.

"*True to Nature and to a fine kind of nature . . . the style is simple an graceful . . . a work of Art, clever and healthy toned.*"—GLOBE.

A 2

THE RECOLLECTIONS OF GEOFFRY HAMLYN.
By Henry Kingsley, Esq.

Second Edition. Crown 8vo. cloth, 6s.

Mr. Henry Kingsley has written a work that keeps up its interest from the first page to the last,—it is full of vigorous stirring life, and though an eager reader may be prompted to skip intervening digressions and details, hurrying on to see what comes of it all, he will, nevertheless, be pretty sure to return and read dutifully all the skipped passages after his main anxiety has been allayed. The descriptions of Australian life in the early colonial days are marked by an unmistakeable touch of reality and personal experience . . . . . Mr. Henry Kingsley has written a book which the public will be more inclined to read than to criticise, and we commend them to each other."—Athenæum.

THE ITALIAN WAR OF 1848-9,
And the Last Italian Poet. By the late Henry Lushington, Chief Secretary to the Government of Malta.
With a Biographical Preface by G. Stovin Venables.

Crown 8vo. cloth, 6s. 6d.

" As the writer warms with his subject, he reaches a very uncommon and characteristic degree of excellence. The narrative becomes lively and graphic, and the language is full of eloquence. Perhaps the most difficult of all literary tasks —the task of giving historical unity, dignity, and interest, to events so recent as to be still encumbered with all the details with which newspapers invest them—has never been more successfully discharged. . . . Mr. Lushington, in a very short compass, shows the true nature and sequence of the event, and gives to the whole story of the struggle and defeat of Italy a degree of unity and dramatic interest which not one newspaper reader in ten thousand ever supposed it to possess."—Saturday Review.

SCOURING OF THE WHITE HORSE.
By the Author of "Tom Brown's School Days."
With numerous Illustrations by Richard Doyle. Eighth Thousand. Imp. 16mo. printed on toned paper, gilt leaves.   8s. 6d.

"The execution is excellent. . . . Like Tom Brown's School Days, the White Horse gives the reader a feeling of gratitude and personal esteem towards the author. The author could not have a better style, nor a better temper, nor a more excellent artist than Mr. Doyle to adorn his book."—Saturday Review,

EDITED BY W. G. CLARK, M.A.
Public Orator in the University of Cambridge.
George Brimley's Essays. With Portrait.

Second Edition. Fcap. 8vo. cloth. 5s.

" One of the most delightful and precious volumes of criticism that has appeared in these days. . . . To every cultivated reader they will disclose the wonderful clearness of perception, the delicacy of feeling, the pure taste, and the remarkably firm and decisive judgment which are the characteristics of all Mr. Brimley's writings on subjects that really penetrated and fully possessed his nature."—Nonconformist.

Cambridge Scrap-Book. Containing in a Pictorial Form a Report on the Manners, Customs, Humours, and Pastimes of the University of Cambridge. Containing nearly 300 Illustrations. Second Edition.   Crown 4to. half-bound, 7s. 6d.

Yes and No; or Glimpses of the Great Conflict.

3 Vols. 1l. 11s. 6d.

"*The best work of its class we have met with for a long time.*"—PATRIOT.

"*Has the stamp of all the higher attributes of authorship.*" — MORNING ADVERTISER.

"*Of singular power.*"—BELL'S MESSENGER.

BY ALEXANDER SMITH,
Author of a "Life Drama, and other Poems."

City Poems.   Fcap. 8vo. cloth, 5s.

"*He has attained at times to a quiet continuity of thought, and sustained strength of coherent utterance . . . he gives us many passages that sound the deeps of feeling, and leave us satisfied with their sweetness.*"—NORTH BRITISH REVIEW.

BY JOHN MALCOLM LUDLOW,
Barrister-at-Law.

British India, its Races, and its History, down to the Mutinies of 1857.   2 vols. fcap. 8vo. cloth, 9s.

"*The best historical Indian manual existing, one that ought to be in the hands of every man who writes, speaks, or votes on the Indian question.*"—EXAMINER.

"*The best elementary work on the History of India.*"—HOMEWARD MAIL.

MEMOIR OF THE REV. GEORGE WAGNER,
Late of St. Stephen's, Brighton.

By J. N. SIMPKINSON, M.A., Rector of Brington,
Northampton.   Second Edition.   Crown 8vo. cloth, 9*s.*

" *A deeply interesting picture of the life of one of a class of men who are indeed the salt of this land.*"—MORNING HERALD.

BY FRANCIS MORSE, M.A.
Incumbent of St. John's, Ladywood, Birmingham.

Working for God.   And other Practical Sermons.
Second Edition.   Fcap. 8vo. 5*s.*

" *For soundness of doctrine, lucidity of style, and above all for their practical teaching, these sermons will commend themselves.*"—JOHN BULL.

" *There is much earnest, practical teaching in this volume.*"—ENGLISH CHURCHMAN.

BY THE REV. J. LLEWELYN DAVIES, M.A.
Rector of Christ Church, St. Marylebone, late Fellow of Trinity College, Cambridge.

The Work of Christ; or the World reconciled to God.
Sermons Preached at Christ Church, St. Marylebone.   With a Preface on the Atonement Controversy.   Fcap. 8vo. cloth, 6*s.*

BY THE REV. D. J. VAUGHAN, M:A.,
Vicar of St. Martin's, Leicester, late Fellow of Trinity College, Cambridge.

Sermons on the Resurrection.   With a Preface.
Fcap. 8vo. cloth, price 3*s.*

CONTENTS :

I.—THE FELLOWSHIP OF CHRIST'S SUFFERINGS.
II.—CHRIST THE RESURRECTION AND THE LIFE.
III.—CHRIST OUR PASSOVER.
IV.—CHRIST THE SHEPHERD.
V.—THE TRUE LIGHT WHICH LIGHTETH EVERY MAN.
VI.—THE CITY OF GOD, AND THE LIGHT THEREOF.
VII.—CHRIST GOING TO THE FATHER, AND THE WAY TO THE FATHER.

BY W. WHEWELL, D.D.
Master of Trinity College, Cambridge.

The Platonic Dialogues for English Readers.
Two Vols. Crown 8vo. cloth, 14s.

"*So readable is this book that no young lady need be deterred from undertaking it: and we are much mistaken if there be not fair readers who will think, as Lady Jane Grey did, that hunting or other female sport is but a shadow compared with the pleasure there is to be found in Plato. The main questions which the Greek master and his disciples discuss are not simply for these in Moral Philosophy schools; they are questions real and practical, which concern Englishmen in public and private life, or their sisters or wives who are busy in lowly and aristocratic households.*"—ATHENÆUM.

THE REPUBLIC OF PLATO.

A New Translation into English. With an Analysis and Notes. By J. LL. DAVIES, M.A., and D. J. VAUGHAN, M.A., Fellows of Trinity College, Cambridge. SECOND EDITION.
8vo. cloth, 10s. 6d.

"*So eloquent and correct a version will, we trust, induce many to become students of the Republic... The whole book is scholarlike and able.*"—GUARDIAN.
"*Free, nervous, idiomatic English, such as will fascinate the reader.*"—NONCON-FORMIST.

BY GEORGE WILSON, M.D., F.R.S.E.,
Regius Professor of Technology in the University of Edinburgh; and Director of the Industrial Museum of Scotland.

Seventh Thousand.

1. The Five Gateways of Knowledge. A Popular Work on the Five Senses.   In fcap. 8vo. cloth, with gilt leaves, 2s. 6d. PEOPLE'S EDITION, in ornamental stiff covers, 1s.

"*Dr. Wilson unites poetic with scientific faculty, and this union gives a charm to all he writes. In the little volume before us he has described the Five Senses in language so popular that a child may comprehend the meaning, so suggestive that philosophers will read it with pleasure.*"—LEADER.

2. The Progress of the Telegraph.   Fcap. 8vo. 1s.

"*Most interesting and instructive... at once scientific and popular, religious and technical; a worthy companion to the 'Gateways of Knowledge.'*"—LITERARY CHURCHMAN.

THE WORKS OF

WILLIAM ARCHER BUTLER, M.A.,

Late Professor of Moral Philosophy in the University of Dublin.

FIVE VOLUMES 8vo. UNIFORMLY PRINTED AND BOUND.

"A man of glowing genius and diversified accomplishments, whose remains fill these five brilliant volumes."—EDINBURGH REVIEW.

SOLD SEPARATELY AS FOLLOWS.

1. Sermons, Doctrinal and Practical. FIRST SERIES. Edited by the Very Rev. THOS. WOODWARD, M.A., Dean of Down. With a Memoir and Portrait. Fifth Edition. 8vo. cloth, 12s.

"Present a richer combination of the qualities for Sermons of the first class than any we have met with in any living writer."—BRITISH QUARTERLY REVIEW.

2. Sermons, Doctrinal and Practical. SECOND SERIES. Edited by J. A. JEREMIE, D.D., Regius Professor of Divinity in the University of Cambridge. Third Edition. 8vo. cloth, 10s. 6d.

"They are marked by the same originality and vigour of expression, the same richness of imagery and illustration, the same large views and catholic spirit, and the same depth and fervour of devotional feeling, which so remarkably distinguished the preceding Series, and which rendered it a most valuable accession to our theological literature."—From DR. JEREMIE'S PREFACE.

3. Letters on Romanism, in Reply to DR. NEWMAN's Essay on Development. Edited by the Very Rev. THOMAS WOODWARD, M.A., Dean of Down. SECOND EDITION. Revised by the Ven. ARCHDEACON HARDWICK. 8vo. cloth, 10s. 6d.

"Deserve to be considered the most remarkable proofs of the Author's indomitable energy and power of concentration."—EDINBURGH REVIEW.

4. Lectures on the History of Ancient Philosophy. Edited from the Author's MSS., with Notes, by WILLIAM HEPWORTH THOMPSON, M.A., Regius Professor of Greek in the University of Cambridge. 2 vols. 8vo., £1 5s.

"Of the dialectic and physics of Plato they are the only exposition at once full, accurate, and popular, with which I am acquainted: being far more accurate than the French, and incomparably more popular than the German treatises on these departments of the Platonic philosophy."—From PROF. THOMPSON'S PREFACE.

THIRD EDITION.

Lectures to Ladies on Practical Subjects. Crown 8vo. 7s. 6d.

By F. D. MAURICE, CHARLES KINGSLEY, J. LL. DAVIES, ARCH-
DEACON ALLEN, DEAN TRENCH, PROFESSOR BREWER, DR. GEORGE
JOHNSON, DR. SIEVEKING, DR. CHAMBERS, F. J. STEPHEN, ESQ. and
TOM TAYLOR, ESQ.

CONTENTS:—Plan of Female Colleges—The College and the Hospital—
The Country Parish—Overwork and Anxiety—Dispensaries—Dis-
trict Visiting—Influence of Occupation on Health—Law as it affects
the Poor—Everyday Work of Ladies—Teaching by Words—Sani-
tary Law—Workhouse Visiting.

" *We scarcely know a volume containing more sterling good sense, or a finer expression of modern intelligence on social subjects.*"—CHAMBERS' JOURNAL.

BY BROOKE FOSS WESTCOTT, M.A.,
Author of "History of the New Testament Canon," &c.

Characteristics of the Gospel Miracles. Sermons preached
before the University of Cambridge. With Notes.
Crown 8vo. cloth, 4s. 6d.

"*An earnest exhibition of important and exalted truth.*"—JOURNAL OF SAC.
LITERATURE.

BY C. A. SWAINSON, M.A.
Principal of the Theological College, and Prebendary of Chichester.

1. The Authority of the New Testament ; the Convic-
tion of Righteousness, and other Lectures delivered before
the University of Cambridge. 8vo. cloth, 12s.

" *These remarkable Lectures deal with most engrossing subjects in an honest and
vigorous spirit. The religious topics which are now uppermost in the mind of
the thoughtful classes among us, and which are fundamental to the Christian,
are here grappled with, we gladly acknowledge, in a courageous, straightfor-
ward way. The reader is led to think healthily and calmly. . . . Our readers
will do well to obtain the book and read it all, there is so much in it of abiding
value.*"—LITERARY CHURCHMAN.

2. The Creeds of the Church. In their Relations to the
Word of God and the Conscience of the Christian. 8vo. cloth, 9s.

3. A Handbook to Butler's Analogy. With a few Notes.
1s. 6d.

BY JULIUS CHARLES HARE, M.A.,

Sometime Archdeacon of Lewes, Rector of Herstmonceux, Chaplain in Ordinary to the Queen, and formerly Fellow and Tutor of Trinity College, Cambridge.

NINE VOLS. 8vo. UNIFORMLY PRINTED AND BOUND.

1. **Charges to the Clergy of the Archdeaconry of Lewes.** During 1840 to 1854, with Notes on the Principal Events affecting the Church during that period. And an Introduction, explanatory of his position in the Church, with reference to the Parties which divide it.

    3 vols. 8vo. cloth, £1 11s. 6d.

2. **Miscellaneous Pamphlets on some of the Leading** Questions agitated in the Church during the years 1845 to 1851.

    8vo. cloth, 12s.

3. **Vindication of Luther against his recent English Assailants.** Second Edition.   8vo. cloth, 7s.

4. **The Mission of the Comforter.** With Notes. Second Edition.   8vo. cloth, 12s.

5. **The Victory of Faith.** Second Edition.   8vo. cloth, 5s.

6. **Parish Sermons.** Second Series.   8vo. cloth, 12s.

7. **Sermons preacht on Particular Occasions.**   8vo. 12s.

The two following books are included among the collected Charges, but are published separately for purchasers of the rest.

Charges to the Clergy of the Archdeaconry of Lewes. Delivered in the years 1843, 1845, 1846. Never before published. With an Introduction, explanatory of his position in the Church, with reference to the Parties that divide it.   8vo. cloth, 6s. 6d.

The Contest with Rome. A Charge, delivered in 1851. With Notes, especially in answer to DR. NEWMAN on the Position of Catholics in England. Second Edition.   8vo. cloth, 10s. 6d.

BY JOHN McLEOD CAMPBELL,

Formerly Minister of Row.

The Nature of the Atonement, and its Relation to Remission of Sins and Eternal Life.

8vo. cloth, 10s. 6d.

" *This is a remarkable book, as indicating the mode in which a devout and intellectual mind has found its way, almost unassisted, out of the extreme Lutheran and Calvinistic views of the Atonement into a healthier atmosphere of doctrine. ... We cannot assent to all the positions laid down by this writer, but he is entitled to be spoken respectfully of, both because of his evident earnestness and reality, and the tender mode in which he deals with the opinions of others from whom he feels compelled to differ.*"—LITERARY CHURCHMAN.

BY THE RIGHT REV. G. E. LYNCH COTTON, D.D.,

Lord Bishop of Calcutta and Metropolitan of India.

Sermons and Addresses delivered in Marlborough College, during Six Years.

Crown 8vo. cloth, price 10s. 6d.

" *We can heartily recommend this volume as a most suitable present for a youth, or for family reading; wherever there are young persons, the teaching of these discourses will be admirable.*"—LITERARY CHURCHMAN.

Sermons: Chiefly connected with Public Events in 1854.

Fcap. 8vo. cloth, 3s.

" *A volume of which we can speak with high admiration.*"

CHRISTIAN REMEMBRANCER.

Charge delivered to the Clergy of Calcutta at his Primary Visitation in September, 1859. 8vo. 2s. 6d.

BY JOHN HAMILTON, Esq. (of St. Ernan's,) M.A.,

St. John's College, Cambridge.

On Truth and Error: Thoughts, in Prose and Verse, on the Principles of Truth, and the Causes and Effects of Error.

Crown 8vo. Cheap Edition, cloth, 5s.

" *A very genuine, thoughtful, and interesting book, the work of a man of honest mind and pure heart; one who has felt the pressure of religious difficulties, who has thought for himself on the matters of which he doubted, and who has patiently and piously worked his way to conclusions which he now reverently but fearlessly utters to the world.*"—NONCONFORMIST.

BY ,CHARLES KINGSLEY, M.A.

Chaplain in Ordinary to the Queen, Rector of Eversley,
and Regius Professor of Modern History in the University of Cambridge.

1. Two Years Ago.   Third Edition.   Crown 8vo. cloth, 6s.

   " Genial, large hearted, humorous, with a quick eye and a keen relish alike
   for what is beautiful in nature and for what is genuine, strong, and earnest in
   man."—GUARDIAN.

2. " Westward Ho!" or the Voyages and Adven-
   tures of Sir Amyas Leigh, Knight, of Borrough, in the County
   of Devon, in the reign of Her most Glorious Majesty Queen
   Elizabeth.   New Edition.   Crown 8vo. cloth, 6s.

   "Almost the best historical novel to our mind of the day."—FRAZER'S
   MAGAZINE.

3. The Heroes: Greek Fairy Tales for my Children.
   New and Cheaper Edition, with Eight Illustrations.   Royal 16mo.
   beautifully printed on toned paper, gilt edges, 5s.

   " We doubt not they will be read by many a youth with an enchained interest
   almost as strong as the links which bound Andromeda to her rock."—BRITISH
   QUARTERLY.

4. Glaucus; or, the Wonders of the Shore.   A Companion for the Sea-side. Containing Coloured Illustrations of the
   Objects mentioned in the Work.   Fourth Edition.   Beautifully
   printed and bound in cloth, gilt leaves.   7s. 6d.

   " Its pages sparkle with life, they open up a thousand sources of unanticipated
   pleasure, and combine amusement with instruction in a very happy and unwonted
   degree."—ECLECTIC REVIEW.

5 Phaethon; or, Loose Thoughts for Loose Thinkers.
   Third Edition.   Crown 8vo. boards, 2s.

6. Alexandria and Her Schools.   Four Lectures delivered
   at the Philosophical Institution, Edinburgh.   With a Preface
                                      Crown 8vo. cloth, 5s.

WORKS BY C. J. VAUGHAN, D.D.

Late Head Master of Harrow School.

1. **Notes for Lectures on Confirmation.** With Suitable Prayers. Third Edition. Fcap. 8vo. limp cloth, red leaves, 1s. 6d.

   "A comprehensive, earnest, and useful manual."—ENGLISH CHURCHMAN.
   "Commends itself at once by its simplicity and by its logical arrangement. Not overloaded with a multitude of points, and it brings those which are introduced before the mind in lucid order and in natural sequence. It will prove, as it is well to be, extensively useful."—THE PRESS.

2. **St. Paul's Epistle to the Romans.** The Greek Text with English Notes. 8vo. cloth, 7s. 6d.

   "For educated young men this commentary seems to fill a gap hitherto unfilled. We find in it a careful elucidation of the meaning of phrases by parallel passages from St. Paul himself, with a nearly continuous paraphrase and explanation by which the very difficult connexion of the argument of the Epistle, with its countless digressions and ellipses and abrupt breaks, is pointedly brought out. An educated lad, who thought for himself, would learn more of the real meaning of St. Paul's words by thoroughly thinking out the suggestive exposition of them here supplied, than by any amount of study bestowed upon more elaborate and erudite works... As a whole, Dr. Vaughan appears to us to have given to the world a valuable book of original and careful and earnest thought bestowed on the accomplishment of a work which will be of much service, and which is much needed."—GUARDIAN.

3. **Memorials of Harrow Sundays.** A Selection of Sermons preached in the School Chapel. With a View of the Interior of the Chapel.

   Second Edition. Crown 8vo. cloth, red leaves, 10s. 6d.

4. **Epiphany, Lent, and Easter.** A Selection of Expository Sermons. Crown 8vo. cloth, red leaves. 7s. 6d.

   "Each exposition has been prepared upon a careful revision of the whole passage ... and the extreme reverence and care with which the author handles Holy Writ, are the highest guarantees of success. Replete with thought, scholarship, earnestness, and all the elements of usefulness."—LITERARY GAZETTE.

5. **Revision of the Liturgy.** Five Discourses. With an Introduction. I. Absolution. II. Regeneration. III. The Athanasian Creed. IV. Burial Service. V. Holy Orders.

   Crown 8vo. cloth, red leaves (1860), 117 pp. 4s. 6d.

   "The large-hearted and philosophical spirit in which Dr. Vaughan has handled the specific doctrines of controversy point him out as eminently fitted to deal with the first principles of the question."—JOHN BULL.

BY THE VENBLE. ARCHDEACON HARDWICK.

Christ and other Masters: A Historical Inquiry into some of the chief Parallelisms and Contrasts between Christianity and the Religious Systems of the Ancient World.

> Part III. Religions of China, America, and Oceanica.
> Part IV. Religions of Egypt and Medo-Persia.
>
> In 8vo. cloth, 7s. 6d. each.

" Never was so difficult and complicated a subject as the history of Pagan religion handled so ably, and at the same time rendered so lucid and attractive." —COLONIAL CHURCH CHRONICLE.

BY THOMAS RAWSON BIRKS, M.A.,

Rector of Kelshall, Examining Chaplain to the Lord Bishop of Carlisle; Author of "The Life of the Rev. E. Bickersteth."

The Difficulties of Belief, in connexion with the Creation and the Fall. Crown 8vo. cloth, 4s. 6d.

" A profound and masterly essay."—ECLECTIC.

" His arguments are original, and carefully and logically elaborated. We may add that they are distinguished by a marked sobriety and reverence for the Word of God."—RECORD.

BY THE VERY REV. R. C. TRENCH, D.D.,

Dean of Westminster.

1. Synonyms of the New Testament.

   Fourth Edition. Fcap. 8vo. cloth, 5s.

2. Hülsean Lectures for 1845—46.

   CONTENTS. 1.—The Fitness of Holy Scripture for unfolding the Spiritual Life of Man. 2.—Christ the Desire of all Nations; or the Unconscious Prophecies of Heathendom.

   Fourth Edition. Fcap. 8vo. cloth, 5s.

3. Sermons Preached before the University of Cambridge. Fcap. 8vo. cloth, 2s. 6d.

BY DAVID MASSON, M.A.,
Professor of English Literature in University College, London.

1. **Life of John Milton, narrated in connexion with the Political, Ecclesiastical, and Literary History of his Time.** Vol. I. 8vo. With Portraits. 18s.

"*Mr. Masson's Life of Milton has many sterling merits . . . his industry is immense; his zeal unflagging; his special knowledge of Milton's life and times extraordinary . . . . with a zeal and industry which we cannot sufficiently commend, he has not only availed himself of the biographical stores collected by his predecessors, but imparted to them an aspect of novelty by his skilful rearrangement.*"—EDINBURGH REVIEW. *April*, 1860.

2. **British Novelists and their Styles: Being a Critical Sketch of the History of British Prose Fiction.** Crown 8vo. cloth, 7s. 6d.

"*A work eminently calculated to win popularity, both by the soundness of its doctrine and the skill of its art.*"—THE PRESS.

3. **Essays, Biographical and Critical: chiefly on English Poets.** 8vo. cloth, 12s. 6d.

CONTENTS.

I. Shakespeare and Goethe.—II. Milton's Youth.—III. The Three Devils: Luther's, Milton's, and Goethe's.—IV. Dryden, and the Literature of the Restoration.— V. Dean Swift.—VI. Chatterton: a Story of the Year 1770.—VII. Wordsworth.—VIII. Scottish Influence on British Literature.—IX. Theories of Poetry.—X. Prose and Verse: De Quincey.

"*Distinguished by a remarkable power of analysis, a clear statement of the actual facts on which speculation is based, and an appropriate beauty of language. These Essays should be popular with serious men.*"—THE ATHENÆUM.

THE ILIAD OF HOMER.
TRANSLATED INTO ENGLISH VERSE.

By I. C. WRIGHT, M.A., Translator of "Dante," late Fellow of Magdalen College, Oxford. Books I.—VI. Crown 8vo. 5s.

"*We know of no edition of the 'sovran poet' from which an English reader can derive on the whole so complete an impression of the immortal Epos.*"— DAILY NEWS.

THE WORKS OF

FREDERICK DENISON MAURICE, M.A.,

Incumbent of St. Peter's, St. Marylebone.

What is Revelation? With Letters on Mr. Mansel's Bampton
Lectures. 10*s.* 6*d.*
Sequel to the Inquiry, "What is Revelation?"
With Letters on Mr. Mansel's Strictures. 6*s.*
Exposition of the Holy Scriptures: .
  (1.) The Patriarchs and Lawgivers. 6*s.*
  (2.) The Prophets and Kings. 10*s.* 6*d.*
  (3.) The Gospel of St. John. 10*s.* 6*d.*
  (4.) The Epistles of St. John. 7*s.* 6*d.*
Exposition of the Ordinary Services of the Prayer
Book: 5*s.* 6*d.*
Ecclesiastical History. 10*s.* 6*d.*
The Doctrine of Sacrifice. 7*s.* 6*d.*
Theological Essays. Second Edition. 10*s.* 6*d.*
The Religions of the World. Third Edition. 5*s.*
Learning and Working. 5*s.*
The Indian Crisis. Five Sermons. 2*s.* 6*d.*
The Sabbath, and other Sermons. 2*s.* 6*d.*
Law on the Fable of the Bees. 4*s.* 6*d.*

The Worship of the Church. A Witness for the
Redemption of the World. 1*s.*
The Word "Eternal" and the Punishment of the
Wicked. Third Edition. 1*s.*
The Name Protestant, and the English Bishopric at
Jerusalem. Second Edition. 3*s.*
The Duty of a Protestant in the Oxford Election. 1847. 1*s.*
The Case of Queen's College, London. 1*s.* 6*d.*
Death and Life. In Memoriam C.B.M. 1*s.*
Administrative Reform. 3*d.*

MANUALS FOR THEOLOGICAL STUDENTS,

UNIFORMLY PRINTED AND BOUND.

This Series of Theological Manuals has been published with the aim of supplying books concise, comprehensive, and accurate, convenient for the Student and yet interesting to the general reader.

.

I.

Introduction to the Study of the Gospels.   By Brooke Foss Westcott, M.A. formerly Fellow of Trinity College, Cambridge.   Crown 8vo. cloth, 10s. 6d.

> "*The worth of Mr. Westcott's volume for the spiritual interpretation of the Gospels is greater than we can readily express even by the most grateful and approving words.   It presents with an unparalleled completeness—the characteristic of the book everywhere being this completeness—wholeness of view, comprehensiveness of representation, the fruits of sacred learning.*"—Non-CONFORMIST.

II.

A General View of the History of the Canon of the New Testament during the FIRST FOUR CENTURIES. By Brooke Foss Westcott, M.A.

Crown 8vo. cloth, 12s. 6d.

> *The Author is one of those who are teaching us that it is possible to rifle the storehouses of German theology, without bearing away the taint of their atmosphere : and to recognise the value of their accumulated treasures, and even track the vagaries of their theoretic ingenuity, without abandoning in the pursuit the clear sight and sound feeling of English common sense . . . . It is by far the best and most complete book of the kind; and we should be glad to see it well placed on the lists of our examining chaplains.*"—GUARDIAN.

> " *Learned, dispassionate, discriminating, worthy of his subject, and the present state of Christian Literature in relation to it.*"—BRITISH QUARTERLY.

> " *To the student in Theology it will prove an admirable Text-Book : and to all others who have any curiosity on the subject it will be satisfactory as one of the most useful and instructive pieces of history which the records of the Church supply.*"—LONDON QUARTERLY.

THEOLOGICAL MANUALS—continued.

III.

History of the Christian Church, during the Middle Ages and the Reformation (A.D. 590–1600).

By the Venerable CHARLES HARDWICK, Archdeacon of Ely.
2 vols. crown 8vo. 10s. 6d. each.

Vol. I. History of the Church to the Excommunication of Luther.
With Four Maps.

Vol. II. History of the Reformation.

Each Volume may be had separately.

"*Full in references and authority, systematic and formal in division, with enough of life in the style to counteract the dryness inseparable from its brevity, and exhibiting the results rather than the principles of investigation.* MR. HARD-WICK *is to be congratulated on the successful achievement of a difficult task.*'
—CHRISTIAN REMEMBRANCER.

"*He has bestowed patient and extensive reading on the collection of his materials; he has selected them with judgment; and he presents them in an equable and compact style.*"—SPECTATOR.

"*To a good method and good materials* MR. HARDWICK *adds that great virtue, a perfectly transparent style. We did not expect to find great literary qualities in such a manual, but we* have *found them; we should be satisfied in this respect with conciseness and intelligibility; but while this book has both, it is also elegant, highly finished, and highly interesting.*"—NONCONFORMIST.

IV.

History of the Book of Common Prayer,

together with a Rationale of the several Offices. By FRANCIS PROCTER, M.A., Vicar of Witton, Norfolk, formerly Fellow of St. Catharine's College, Cambridge. Fourth Edition, revised and enlarged. Crown 8vo. cloth, 10s. 6d.

"MR. PROCTER'S 'History of the Book of Common Prayer' *is by far the best commentary extant . . . . . . Not only do the present illustrations embrace the whole range of original sources indicated by* MR. PALMER, *but* MR. PROCTER *compares the present Book of Common Prayer with the Scotch and American forms; and he frequently sets out in full the Sarum Offices. As a manual of extensive information, historical and ritual, imbued with sound Church princi-ples, we are entirely satisfied with* MR. PROCTER's *important volume.*"
CHRISTIAN REMEMBRANCER.

"*It is indeed a complete and fairly-written history of the Liturgy; and from the dispassionate way in which disputed points are touched on, will prove to many troubled consciences what ought to be known to them, viz.:—that they may, without fear of compromising the principles of evangelical truth, give their assent and consent to the contents of the Book of Common Prayer.* MR. PROCTER *has done a great service to the Church by this admirable digest.*"
CHURCH OF ENGLAND QUARTERLY.

MR. CORNWALL SIMEON'S
Stray Notes on Fishing and Natural History, With Illustrations. 7s. 6d.

"*Written in a hearty and sportsmanlike spirit, breathing freshly of the river side and abounding in quaint and piquant anecdote . . . . sound practical information, at once profitable to the tyro and entertaining to the proficient.*"—LITERARY GAZETTE.

"*As pleasant a volume of its kind as any that has appeared since 'White's History of Selborne.*"—JOHN BULL.

"*Excellent and thoroughly practical, just what the amateur needs.*"—ERA.

"*If this remarkably agreeable work does not rival in popularity the celebrated 'White's Selborne,' it will not be because it does not deserve it . . . the mind is almost satiated with a repletion of strange facts and good things.*"—FIELD, July 28, 1860.

"*What is to be said is said briefly and well . . . one of the most sensible and amusing of a class of books welcome always.*"—EXAMINER, Sept. 8, 1860.

MR. WESTLAND MARSTON'S NOVEL
"A Lady in Her Own Right." 10s. 6d.

"*A perfect masterpiece of chaste and delicate conception, couched in spirited and eloquent language, abounding in poetical fancies. . . . Seldom have we met with anything more beautiful, perfect, or fascinating than the heroine of this work.*" LEADER.

Artist and Craftsman. A Novel. 10s. 6d.

"*There are many beauties which we might have pointed out, but we prefer counselling our readers to read the book and discover for themselves.*"—LITERARY GAZETTE.

Blanche Lisle, and other Poems. By CECIL HOME. 4s. 6d.

"*The writer has music and meaning in his lines and stanzas, which, in the selection of diction and gracefulness of cadence, have seldom been excelled.*"—LEADER, June 2, 1860.

"*Far above most of the fugitive poetry which it is our lot to review . . . full of a true poet's imagination.*"—JOHN BULL.

MACMILLAN AND CO.'S

Class Books for Colleges and Schools.

I. ARITHMETIC AND ALGEBRA.

Arithmetic. For the use of Schools. By BARNARD SMITH, M.A.
New Edition (1860). 348 pp. Answers to all the Questions. Crown 8vo. 4*s.* 6*d.*

Key to the above. Crown 8vo. 8*s.* 6*d.* Second Edition
thoroughly Revised. [*In the Press.*

Arithmetic and Algebra in their Principles and Applications.
With numerous Examples, systematically arranged. By BARNARD SMITH, M.A.
Seventh Edition (1860), 696 pp. Crown 8vo. 10*s.* 6*d.*

Exercises in Arithmetic. By BARNARD SMITH, M.A. Part I.
48 pp. (1860). Crown 8vo. 1*s.* [*Part II. Nearly ready.*

Arithmetic in Theory and Practice. For Advanced Pupils. By
J. BROOK SMITH, M.A. Part First. 164 pp. (1860). Crown 8vo. 3*s.* 6*d.*

A Short Manual of Arithmetic. By C. W. UNDERWOOD, M.A.
96 pp. (1860). Fcp. 8vo. 2*s.* 6*d.*

Algebra. For the use of Colleges and Schools. By I. TODHUNTER,
M.A. Second Edition. Crown 8vo. 516 pp. (1860). 7*s.* 6*d.*

II. TRIGONOMETRY.

Introduction to Plane Trigonometry. For the use of Schools.
By J. C. SNOWBALL, M.A. Second Edition (1847). 8vo. 5*s.*

Plane Trigonometry. For Schools and Colleges. By I. TODHUNTER,
M.A. 272 pp. (1859). Crown 8vo. 5*s.*

Spherical Trigonometry. For Colleges and Schools. By I.
TODHUNTER, M.A. 112 pp. (1859). Crown 8vo. 4*s.* 6*d.*

Plane Trigonometry. With a numerous Collection of Examples.
By R. D. BEASLEY, M.A. 106 pp. (1858). Crown 8vo. 3*s.* 6*d.*

Plane and Spherical Trigonometry. With the Construction and
Use of Tables of Logarithms. By J. C. SNOWBALL, M.A. Ninth Edition, 240 pp.
(1857). Crown 8vo. 7*s.* 6*d.*

III. MECHANICS AND HYDROSTATICS.

Elementary Treatise on Mechanics. With a Collection of
Examples. By S. PARKINSON, B.D. Second Edition. [*In the Press.*

Elementary Course of Mechanics and Hydrostatics. By J. C.
SNOWBALL, M.A. Fourth Edition. 110 pp. (1851). Crown 8vo. 5*s.*

MECHANICS AND HYDROSTATICS—*continued*.

Elementary Hydrostatics. With numerous Examples and
Solutions. By J. B. Phear, M.A. Second Edition. 156 pp. (1857). Crown 8vo.
5*s*. 6*d*.

Analytical Statics. With numerous Examples. By I. Todhunter,
M.A. Second Edition. 330 pp. (1858). Crown 8vo. 10*s*. 6*d*.

Dynamics of a Particle. With numerous Examples. By P. G.
Tait, M.A. and W. J. Steele, M.A. 304 pp. (1856). Crown 8vo. 10*s*. 6*d*.

A Treatise on Dynamics. By W. P. Wilson, M.A. 176 pp.
(1850). 8vo. 9*s*. 6*d*.

IV. ASTRONOMY AND OPTICS.

Plane Astronomy. Including Explanations of Celestial Phenc-
mena and Instruments. By A. R. Grant, M.A. 128 pp. (1850). 8vo. 6*s*.

Elementary Treatise on the Lunar Theory. By H. Godfray,
M.A. Second Edition. 119 pp. (1859). Crown 8vo. 5*s*. 6*d*.

A Treatise on Optics. By S. Parkinson, B.D. 304 pp. (1859).
Crown 8vo. 10*s*. 6*d*.

V. GEOMETRY AND CONIC SECTIONS.

Geometrical Treatise on Conic Sections. With a Collection of
Examples. By W. H. Drew, M.A. 121 pp. (1857). 4*s*. 6*d*.

**Plane Co-ordinate Geometry as applied to the Straight Line and
the Conic Sections.** By I. Todhunter, M.A. Second Edition. 316 pp. (1858).
Crown 8vo. 10*s*. 6*d*.

Elementary Treatise on Conic Sections and Algebraic Geometry.
By G. H. Puckle, M.A. Second Edition. 264 pp. (1856). Crown 8vo. 7*s*. 6*d*.

Examples of Analytical Geometry of Three Dimensions. With
the Results. Collected by I. Todhunter, M.A. 76 pp. (1858). Crown 8vo. 4*s*.

VI. DIFFERENTIAL AND INTEGRAL CALCULUS.

The Differential Calculus. With numerous Examples. By I.
Todhunter, M.A. Third Edition. 404 pp. (1860). Crown 8vo. 10*s*. 6*d*.

The Integral Calculus, and its Applications. With numerous
Examples. By I. Todhunter, M.A. 268 pp. (1857). Crown 8vo. 10*s*. 6*d*.

A Treatise on Differential Equations. By George Boole, D.C.L.
486 pp. (1859). Crown 8vo. 14*s*.

A Treatise on the Calculus of Finite Differences. By George
Boole, D.C.L. 248 pp. (1860). Crown 8vo. 10*s*. 6*d*.

VII. PROBLEMS AND EXAMPLES.

A Collection of Mathematical Problems and Examples. With
Answers. By H. A. Morgan, M.A. 190 pp. (1858). Crown 8vo. 6s. 6d.

Senate-House Mathematical Problems. With Solutions—

1848-51. By FERRERS and JACKSON. 8vo. 15s. 6d.
1348-51. (Riders.) By JAMESON. 8vo. 7s. 6d.
1854. By WALTON and MACKENZIE. 8vo. 10s. 6d.
1857. By CAMPION and WALTON. 8vo. 8s. 6d.
1860. By ROUTH and WATSON. Crown 8vo. 7s. 6d.

VIII. LATIN.

Help to Latin Grammar ; or, the Form and Use of Words in
Latin. With Progressive Exercises. By Josiah Wright. M.A. 175 pp. (1855).
Crown 8vo. 4s. 6d.

The Seven Kings of Rome. A First Latin Reading Book. By
Josiah Wright, M.A. Second Edition. 138 pp. (1857). Fcap. 8vo. 3s,

Vocabulary and Exercises on "The Seven Kings." By Josiah
Wright, M.A. 94 pp. (1857). Fcap. 8vo. 2s. 6d.

A First Latin Construing Book. By E. Thring, M.A. 104 pp.
(1855). Fcap. 8vo. 2s. 6d.

Rules for the Quantity of Syllables in Latin. 10 pp. (1858).
Crown 8vo. 1s.

Theory of Conditional Sentences in Latin and Greek. By R.
Horton Smith, M.A. 30 pp. (1859). 8vo. 2s. 6d.

Sallust.—Catilina and Jugurtha. With English Notes. For
Schools. By Charles Merivale. B.D. Second Edition, 172 pp. (1858). Fcap.
8vo. 4s. 6d.
    Catilina and Jugurtha may be had separately, price 2s. 6d. each.

Juvenal. For Schools. With English Notes and an Index. By
J. E. Mayor, M.A. 464 pp. (1853). Crown 8vo. 10s. 6d.

IX. GREEK.

Hellenica ; a First Greek Reading Book. Being a History of
Greece, taken from Diodorus and Thucydides. By Josiah Wright. M.A. Second
Edition. 150 pp. (1857). Fcap. 8vo. 3s. 6d.

Demosthenes on the Crown. With English Notes. By B.
Drake, M.A. Second Edition, to which is prefixed Æschines against Ctesiphon.
With English Notes. (1860). Fcap. 8vo 5s.

Demosthenes on the Crown. Translated by J. P. Norris, M.A.
(1850). Crown 8vo. 3s.

GREEK—*continued.*

Thucydides. Book VI. With English Notes and an Index.
By P. Frost, Jun. M.A. 110 pp. (1854). 8vo. 7s. 6d.

Æschylus. The Eumenides. With English Notes and Translation. By B. Drake, M.A. 144 pp. (1853). 8vo. 7s. 6d.

St. Paul's Epistle to the Romans : With Notes. By Charles John Vaughan, D.D. 157 pp. (1859). 8vo. 7s. 6d.

X. ENGLISH GRAMMAR.

The Child's English Grammar. By E. Thring, M.A. Demy 18mo. New Edition. (1857). 1s.

Elements of Grammar taught in English. By E. Thring, M.A. Third Edition. 136 pp. (1860). Demy 18mo. 2s.

Materials for a Grammar of the Modern English Language. By G. H. Parminter, M.A. 220 pp. (1856). Fcap. 8vo. 3s. 6d.

XI. RELIGIOUS.

History of the Christian Church during the Middle Ages. By Archdeacon Hardwick. 482 pp. (1853). With Maps. Crown 8vo. cloth. 10s. 6d.

History of the Christian Church during the Reformation. By Archdeacon Hardwick. 459 pp. (1850). Crown 8vo. cloth. 10s. 6d.

History of the Book of Common Prayer. By Francis Procter, M.A. 464 pp. (1860). Fourth Edition. Crown 8vo. cloth. 10s. 6d.

History of the Canon of the New Testament during the First Four Centuries. By Brook Foss Westcott, M.A. 594 pp. (1855). Crown 8vo. cloth. 12s. 6d.

Introduction to the Study of the Gospels. By Brooke Foss Westcott, M.A. (1860). Crown 8vo. cloth. 10s. 6d.

The Church Catechism Illustrated and Explained. By Arthur Ramsay, M.A. 204 pp. (1854). 18mo. cloth. 3s. 6d.

Notes for Lectures on Confirmation : With Suitable Prayers. By C. J. Vaughan, D.D. Third Edition. 70 pp. (1859). Fcap. 8vo. 1s. 6d.

Hand-Book to Butler's Analogy. By C. A. Swainson, M.A. 55 pp. (1856). Crown 8vo. 1s. 6d..

History of the Christian Church during the First Three Cen- turies, and the Reformation in England. By William Simpson, M.A. 307 pp. (1857). Fcap. 8vo. cloth. 5s.

Analysis of Paley's Evidences of Christianity. By Charles H. Crosse, M.A. 115 pp. (1855). 18mo. 3s. 6d.

ANNOUNCEMENTS.

I.

On the Origin and Succession of Life on the Earth.
By JOHN PHILLIPS, M.A. F.R.S. F.G.S., Reader in Geology in
the University of Oxford, and Rede Lecturer in the University
of Cambridge, &c. With Illustrations.   Crown 8vo.

II.

Life of Edward Forbes, THE NATURALIST. By GEORGE
WILSON, M.D., late Professor of Technology in the University of
Edinburgh, and ARCHIBALD GEIKIE, F.G.S., of the Geologica
Survey.

III.

Introduction to the Study and Use of the Psalms.
By the Rev. J. F. THRUPP, Author of "An Investigation intc
the Topography, &c. of Ancient Jerusalem," &c.   2 vols. 8vo.

IV.

Rays of Sunlight for Dark Days : A Book of Selections
for the Suffering.   Royal 32mo. elegantly printed,

V.

An Elementary Treatise on Quaternions. With numerous
Examples. By P. G. TAIT, M.A., Professor of Natural Philosophy
in the University of Edinburgh.

VI.

A Treatise on the Dynamics of a Rigid Body. By
E. J. ROUTH, M.A., and C. A. SMITH, B.A., Fellows of St. Peter's
College, Cambridge.

VII.

A Treatise on Geometry of Three Dimensions. By
PERCIVAL FROST, M.A., St. John's College, and JOSEPH WOLSTEN-
HOLME, M.A., Christ's College, Cambridge.

VIII.

An Elementary Treatise on Statics. By GEORGE RAWLIN-
SON, M.A., late Professor of Natural Philosophy in Elphinstone
Institution, Bombay, formerly of Emmanuel College, Cambridge.

IX.

A Treatise on Trilinear Co-ordinates. By N. M. FERRERS,
M.A., Fellow and Mathematical Lecturer of Gonville and Caius
College.

R. CLAY, PRINTER, BREAD STREET HILL.

www.ingramcontent.com/pod-product-compliance
Lightning Source LLC
Chambersburg PA
CBHW021827190326
41518CB00007B/772